GROW IT! YIELD IT!

Everything on Planting Strawberries
for Beginner's Success

John Klein

Absolute Author
Publishing House

Grow it! Yield it! Everything on Planting Strawberries for Beginner's Success
Copyright © 2020 by John Klein
All rights reserved.

Publisher: Absolute Author Publishing House
Editor: Dr. Melissa Caudle
Junior Editor: Paul S. Dupre
Cover Designer: John Klein
Photos: Stock Nation. Use with permission.

Paperback ISBN: 978-1-64953-162-9
eBook ISBN: 978-1-64953-163-6

DEDICATION

To all those who love eating from Mother Earth.

Table of Contents

Preface

This guide is for anyone adventurous, loving to try new things, whether gardening for fun or for improving health. You'll learn everything to grow strawberries successfully from learning how to choose which variety you should grow to harvest that first juicy red fruit! Join me on this journey as we embark on an exciting learning experience!

John Klein

Chapter 1

Why you Should Grow Strawberries

Health Benefits

Strawberries (scientifically known as Fragaria ×
ananassa) are known as a delicious, highly concentrated
nutrient-rich fruit. Fruit from this plant is rich in vitamin
C, which is crucial for immune support. They are also
jam-packed with loads of antioxidants, which have been
shown to help with certain cancers and reversing the
aging process! Strawberries contain the most vitamin C
out of all North American fruits, excluding tropical fruits.
These plants even have the power to fight allergies, as
stawberryplants.org states, "Strawberries are a
wonderful source of a plant chemical called quercetin."
This chemical does the following "Quercetin is a natural
antihistamine. It stops the release of inflammatory

compounds from mast cells in the nose and throat. It blocks the action of hyaluronidase, which breaks down the walls of capillaries, so they leak and cause swelling. When capillaries don't break down, mast cells don't release their histamine, and allergy symptoms don't occur." (*stawberryplants.org*) Wow, what excellent benefits to help ease specific symptomatic issues! Strawberries even help with the absorption of Vitamin C into the body by containing certain easy to digest minerals which make Vitamin C more absorbable.

Here Are Some Fun Facts

- The strawberry fruit was mentioned in ancient Roman literature regarding its medicinal use.
- The first known extensive, man-made propagation of wild strawberries was from the French, and no other than King Charles V, king from 1364 to

1380. He had 1,200 wild strawberry plants in his royal garden.

- Did you know that the first garden strawberry varieties, as we know them today, have been propagated by people since the 1750s? The first known man-bred variety came from Brittany, France.
- Strawberries are the only fruits to have their seeds on the outside of the fruit.
- Strawberries are actually members of the Rose family.
- Plants are perennial (meaning they come back every year).
- Americans eat an average of 3.5 pounds of strawberries annually.

Supplement Your Diet

Growing your own food doesn't have to be hard. In this guide, you'll find it's quite easy if you follow what I've laid out. Have you noticed that whenever you've had something home-grown straight out of your own garden or from a friend's, it just tastes better and more nutritious? Likely, your fruit probably is more nutritious due to the following reasons. Most commercial grown products can't compete with natural home-grown fruits and veggies due to lacking vitamin and mineral content. In commercial operations, most of the fruit is grown on a big scale with lots of sprays and or chemical fertilizers, thus lacking in good organic composted soil that feeds your plants and fruit, which gives it that extra yummy nutrient content. Realize that this is an industry, and they have to feed many people, though for your purposes, maybe try your hand at growing some of your own food more naturally! Simply growing some of your own strawberries can save you hundreds of dollars in a few short years, possibly more, depending on your consumption. I say help yourself, and your family, by growing higher-quality, more nutritious, less expensive fruit, which can also make you proud!

Spend More Time Outdoors

Do you remember playing outside on those long summer nights when you were younger? Remember all the fresh air and sunlight you got, maybe even a few sunburns. Well, with a little sunscreen and TLC, chances are you were probably alright; nevertheless, our bodies need Vitamin D. This key nutrient is one of the most lacking nutrients in American society alone! Vitamin D strengthens our bones, allowing better calcium absorption from our foods, and even aiding our immune systems. Gardening and just growing your own food will not only provide you the obvious benefit of lowering costs of food, but it will also allow you to reap the benefits of the great outdoors even if you live in the city, and allow your bodies a crucial vitamin that is severely lacking in society today. So, I say help your family and yourself out and start a new tradition!

Spend Time with Family, Start a Tradition

Have you always wanted to get outside more? Or maybe you've wanted to give your kids something that you never did growing up. Regardless, isn't it nice to think back to our childhoods and remember fond memories of having a garden in the summer, having fresh fruits and veggies at a moment's notice? Tell me you can't get any fresher than going out your door and harvesting something in your back-yard. I can recall having a garden every summer. My family was big on that. It supplied us with something to do as kids, and my Father and Mother enjoyed having one. Not to mention having one saved us hundreds, maybe thousands of dollars over the years in grocery bills, and supplied us with much healthier foods overall. I want you and your family to have a chance to make fond memories like I did and learn a valuable fading skill in our society today when most people have a tough time growing a plant on their window sill! Let's get back to our roots and mother nature and get to growing!

Chapter 2
Choosing Between Different Varieties

What to Look For

When choosing between different varieties of strawberries, you must look at how you are trying to use the fruit? Are you using them in smoothies? Are you using them for baking pies? Are you using them to freeze and enjoy later? We must assess all these things when choosing your varieties. Some varieties will do better in colder climates, most in moderate climates.

Some will do better in warmer climates, so check out your climate zoning and select suitable varieties based on the information below for your environment.

A fantastic reference to use is this website *strawberryplants.org/strawberry-varieties*. You'll find everything to include dozens of varieties to choose from based on all the above. Also, at *strawberryplants.org*, you'll be able to search what does best in your USDA Hardiness Zone based on entering your state and/or location.

In this guide, you'll find that I've included information on some of the best varieties to grow based on individual states in the U.S. and territories in Canada. This works for everyone in the U.S. and Canada; however, other countries have similar resources available. I grew up with the Hood strawberry varieties back in Oregon. They did amazing, though everywhere else around the country is different, so please use this resource and the following information provided in this guide to find varieties you'd be able to grow based on your location.

Taste

Don't forget this part! It's one of the most crucial when growing to consume. The varieties that grow in your area can be numerous; therefore, I recommend skipping to your state or territory in this reference part of the book, then using your nearby garden center or this great online resource, *strawberryplants.org*, to further inform you on tastes of each variety listed before purchasing a bunch of plants. You'll be able to see the ripening times for the varieties I'll list below, as well as the taste of the fruit. The varieties are too many to include all the fine details here, especially considering there are 103 different distinct species and subspecies of strawberries. (According to the U.S. Department of Agriculture). If I added all these in this guide, this would be an encyclopedia length and not a simple yet complete guide to getting you outside growing like a pro as soon as possible!

Variety

It's always best when growing fruit-bearing plants to plant a few varieties to see which one will do the best in your specific micro-climate. I say this because 10 miles away, one variety may do better than another. It just depends on light, soil, water, and care of plants, plus many other micro-cultures that play into favorability. Growing a few different varieties will make for much better cross-pollination between your plants and produce the highest fruit production. Stick with these tricks and follow up with suitable varieties for your area, and you'll win almost every time!

Use

Depending on how you intend to use your plants should dictate the varieties you choose to grow. Factors to consider here are:

- Jam/Jelly use
- Eating as is
- Freezing (some keep better)
- Deserts/smoothies/Wine use
- Multipurpose use

Varieties to Check Out by State

(All are recommended by local County Cooperative Extension Offices in localities/ Universities/ or *strawberry plants.org* database).

1. Alabama

Albritton, Allstar, Cardinal, Chandler, Delite, Douglas, Earlibelle, Earliglow, Sunrise. (According to the Alabama Cooperative Extension Service of Alabama A&M and Auburn Universities).

2. Alaska

Brighton, Fern, Hecker, Irvine, Mrak, Muir, Ogallala, Ozark Beauty, Quinault, Selva, Streamliner, Superfection, Tillicum, Tribute, Tristar, Yolo. (According to the University of Alaska Fairbanks Cooperative Extension Service).

3. Arizona

Camarosa, Chandler. (According to the University of Arizona Citrus Agricultural Center). Arizona is not a great location to grow strawberries, mainly due to heat and desert conditions. However, it still may be done well if proper care is followed, which you'll find in this guide. ☺

4. Arkansas

Cardinal, Camarosa, Chandler, Delmarvel, Earliglow, Lateglow, Noreaster, Sweet Charlie, Tribute, Tristar. (According to the University of Arkansas Department of Agriculture Cooperative Extension Service).

5. California

Albion, Aromas, Camarosa, Camino Real, Chandler, Diamante, Gaviota, Oso Grande, Pacific, Seascape, Selva, Ventana. (According to the California Strawberry Commission).

6. Colorado

Catskill, Empire, Fairfax, Fort Laramie, Geneva, Guardian, Marlate, Ogallala, Ozark Beauty, Quinault, Redchief, Red Rich, Redstar, Robinson, Superfection, Tribute. (Colorado State University Cooperative Extension Service).

7. Connecticut

Brunswick, Cabot, Clancy, Darselect, Earliglow, Eros, Honeoye, Jewel, L'Amour, Sable. (According to the New England Vegetable and Fruit Conference).

8. Delaware

Allstar, Delite, Earliglow, Guardian, Late Glow, Red Chief, Sparkle, Tribute, Tristar. (According to the University of Delaware College of Agriculture & Natural Resources Cooperative Extension).

9. Florida

Calibrate, Camarosa, Florida Belle, Florida 90, Rosa Linda, Sequoia, Sweet Charlie, Strawberry Festival, Tioga. (According to the University of Florida University Relations Department).

10. Georgia

Apollo, Delite, Cardinal, Earliglow, Sunrise, Surecrop. (According to the University of Georgia College of Agricultural & Environmental Sciences).

11. Hawaii

Eversweet, Quinault, Seascape. Although strawberries are grown commercially on the Islands, and the *Fragaria chiloensis* species of strawberries grow at elevation, they are more difficult to grow in the tropical environment due mainly to fungus diseases and not highly recommended. However, they are possible to grow with a little more TLC. The three varieties listed are sold in nurseries on the main island of Hawaii.

12. Idaho

Allstar, Benton, Blomidon, Catskill, Cavendish, Earliglow, Fort Lamarie, Gloopscap, Guardian, Honeoye, Jewel, Lateglow, Lester, Micmac, Quinault, Redchief, Scott, Shuksan, Surecrop, Totem, Tribute, Tristar. (According to the University of Idaho Extension Service).

13. Illinois

Allstar, Annapolis, Delmarvel, Earliglow, Honeoye, Jewel, Kent, Seneca, Tribute, Tristar. (According to the University of Illinois Extension Service).

14. Indiana

Delite, Earliglow, Fort Laramie, Guardian, Sunrise, Ozark Beauty, Redchief, Sparkle, Surecrop. (According to the Purdue University Extension Service).

15. Iowa

Annapolis, Cavendish, Delmarvel, Honeoye, Jewel, Kent, Mohawk, Primetime, Winona. (According to the Iowa State University Southeast Research and Demonstration Farm).

16. Kansas

Allstar, Earliglow, Guardian, Northeaster, Ogallala, Ozark Beauty, Primetime, Redchief, Tribute, Tristar. (According to the Kansas State University Agricultural Experiment Station and Cooperative Extension Service's Horticultural Report).

17. Kentucky

Camarosa, Chandler, Jewel, Northeaster, Sweet Charlie. (According to the University of Kentucky Department of Horticulture and Landscape Architecture's Fruit and Vegetable Crops Research Report).

18. Louisiana

Camarosa, Camino Real, Strawberry Festival. (According to the Louisiana State University Agriculture Center Research & Extension).

19. Maine

Allstar, Bounty, Catskill, Earliglow, Guardian, Lateglow, Midway, Mira, Mohawk, Northeaster, Surecrop. (According to the University of Maine Cooperative Extension Service).

20. Maryland

Allstar, Bish, Chandler, Darselect, Eros, Jewel, KRS-10, Oviation, Seascape. (According to the University of Maryland Agricultural Experiment Station). Flavorfest (recommended by Kim Lewer's of the USDA's Agricultural Research Service).

21. Massachusetts

Catskill, Earlidawn, Fletcher, Guardian, Midway, Raritan, Redchief, Sparkle, Surecrop. (According to farminfo.org).

22. Michigan

Allstar, Annapolis, Bounty, Cavendish, Chambly, Delmarvel, Earliglow, Glooscap, Honeoye, Jewel, Redchief, Tribute, Tristar. (According to the Michigan State University Extension Van Buren County).

23. Minnesota

Cavendish, Kent, Mesabi, Winona. (According to the University of Minnesota Agricultural Experiment Station and Extension Service).

24. Mississippi

Cardinal, Chandler, Comet, Dixieland, Douglas, Florida 90, Pocahontas, Sunrise, Tangi, Tennessee Beauty. (According to the Mississippi State University Extension Service).

25. Missouri

Allstar, Cardinal, Earliglow, Guardian, Honeoye, Jewel, Lateglow, Ogallala, Ozark Beauty, Redchief,

Sparkle, Surecrop, Tribute, Tristar. (According to the University of Missouri Horticultural M.U. Guide).

26. Montana

Catskill, Fern, Fort Laramie, Gem, Glooscap, Hecker, Honeoye, Ogallala, Red Rich, Redcoat, Senator Dunlap, Sparkle, Streamliner, Tribute, Tristar, Veestar, Vibrant. (According to the Montana State University Extension Service).

27. Nebraska

Earliglow, Ft. Laramie, Ogallala, Sunrise, Surecrop, Redchief, Tribute, Tristar. (According to the University of Nebraska Lincoln Extension in Lancaster County).

28. Nevada

Camarosa, Chandler. Note: Nevada is not considered a suitable location for strawberry cultivation. (Mainly because of the climate and environment). With extra care, they'll grow and do well.

29. New Hampshire

Allstar, Cavendish, Cornwallis, Earliglow, Redchief, Sparkle. (According to the University of New Hampshire Cooperative Extension).

30. New Jersey

Delmarvel, Earliglow, Guardian, Latestar, Lester, Northeaster, Raritan, Redchief, Sparkle, Tribute, Tristar. (According to the National Sustainable Agriculture Information Service).

31. New Mexico

Fern, Fort Laramie, Gem, Guardian, Ogallala, Ozark Beauty, Quinault, Robinson, Selva, Sequoia, Streamliner, Superfection, Surecrop, Tribute, Tristar, Tufts. (According to the New Mexico State University Cooperative Extension Service and College of Agriculture and Home Economics).

32. New York

Allstar, Bounty, Cavendish, Delite, Earliglow, Fletcher, Guardian, Honeoye, Jewel, Kent, Raritan, Redchief, Scott. (According to the Cornell Cooperative Extension Suffolk County).

33. North Carolina

Albion, Bish, Camarosa, Camino Real, Chandler, Gaviota, Gem Star, Oso Grande, Seascape, Strawberry Festival, Sweet Charlie, Treasure, Ventana. (According to the North Carolina Strawberry Association).

34. North Dakota

Dunlap, Ft. Laramie, Gem, Honeoye, Redcoat, Stoplight, Trumpeter. (According to the North Dakota State Agricultural and University Extension).

35. Ohio

Delite, Earliglow, Guardian, Kent, Lateglow, Lester, Midway, Redchief, Surecrop, Tribute, Tristar. (According to the Ohio State University Extension).

36. Oklahoma

Albritton, Allstar, Apollo, Arking, Blakemore, Canoga, Cardinal, Chandler, Delite, Earliglow, Fletcher, Guardian, Holiday, Hood, Lateglow, Luscious Lady, Ozark Beauty, Scott, Spring Giant, Sunrise, Surecrop, Tennessee Beauty, Trumpeter. (According to the Oklahoma State University Cooperative Extension Service).

37. Oregon

Benton, Fern, Ft. Laramie, Hecker, Hood, Olympus, Ozark Beauty, Puget Reliance, Quinault, Rainier, Redcrest, Selva, Shuksan, Sumas, Tillikum, Tristar, Totem. (According to the Oregon State University Extension Service).

38. Pennsylvania

Albion, Allstar, Camarosa, Chandler, Darselect, Earliglow, Everest, Evie-2, Honeoye, Jewel, L'Amour, Seascape, Sweet Charlie, Tribute, Tristar, Wendy. (According to the Penn State University Small-scale and Part-time Farming Project).

39. Rhode Island

Brunswick, Cabot, Clancy, Darselect, Earliglow, Eros, Honeoye, Jewel, L'Amour, Sable. (According to the New England Vegetable and Fruit Conference).

40. South Carolina

Albritton, Apollo, Cardinal, Chandler, Delite, Douglas, Earliglow, Florida 90, Sunrise, Surecrop, Tioga. (According to the Clemson University Cooperative Extension Service).

41. South Dakota

Annapolis, Bounty, Crimson King, Earliglow, Ft. Laramie, Glooscap, Honeoye, Jewel, Kent, Ogallala, Ozark Beauty, Redcoat, Selva, Seneca, Settler, Sparkle, Tribute, Tristar, Trumpeter, Veestar. (According to the South Dakota State University Cooperative Extension Service).

42. Tennessee

Allstar, Cardinal, Delite, Delmarvel, Earliglow, Guardian, Lateglow, Red Chief, Scott, Surecrop, Tribute, Tristar. (According to the Agricultural Extension Service of the University of Tennessee).

43. Texas

Allstar, Cardinal, Chandler, Douglas, Pajaro, Sequoia. (According to the Texas A&M System, Department of Horticultural Sciences, AgriLife Extension).

44. Utah

Allstar, Chandler, Earliglow, Evie-2, Honeoye, Jewel, Ogallala, Seascape, Sparkle, Tribute. (According to the Utah State University Cooperative Extension).

45. Vermont

Allstar, Annapolis, Brunswick, Cabot, Cavendish, Clancey, Cornwallis, Darselect, Earliglow, Everest, Honeoye, Jewel, Kent, L'Amour, Lateglow, Mesabi, Mic Mac, Mira, Mohawk, Northeaster, Sable, Seascape, Seneca, Sparkle, Tribute, Tristar, Veestar, Winona. (According to the University of Vermont Extension).

46. Virginia

Allstar, Delite, Delmarvel, Earliglow, Honeoye, Lateglow, Ozark Beauty, Redchief, Sunrise, Surecrop, Tribute, Tristar. (According to the Virginia Cooperative Extension).

47. Washington State

Hood, Nanaimo, Puget Reliance, Quinault Rainier, Selva, Shuksan, Tillicum, Totem, Tribute, Tristar. (According to the Washington State University Extension).

48. West Virginia

Allstar, Annapolis, Earliglow, Sable, Seneca, Surecrop. (According to the West Virginia University Extension Service).

49. Wisconsin

Annapolis, Cavendish, Crimson Fern, Fort Laramie, King, Earliglow, Glooscap, Honeoye, Jewel, Kent, Lateglow, Lester, Mesabi, Mira, Ogallala, Ozark Beauty, Raritan, Redchief, Seascape, Selva, Seneca, Sparkle, Tribute, Tristar, Winona. (According to the Cooperative Extension System of the University of Wisconsin).

50. Wyoming

Dunlap, Fort Laramie, Guardian, Honeoye, Ogallala, Ozark Beauty, Quinault, Redcoat, Surecrop, Tribute, Tristar, Trumpeter. (According to the University of Wyoming College of Agriculture).
(Courtesy of *strawberryplants.org*)

Strawberry Varieties to Grow in Canada by Providence

1. Alberta

(As recommended by Alberta Agriculture and Rural Development).
June-bearing: Bounty, Cavendish, Glooscap, Honeoye, Kent; Everbearing: Fort Laramie, Ogallala; Day-neutral: Albion, Fern, Seascape, Tristar.

2. BC British Columbia

(As recommended by the British Columbia Ministry of Agriculture; Pacific Agri-Food Research Centre, Agriculture, and Agri-Food Canada).

June-bearing: Clancy, Hood, Honeoye, Nisgaa (B.C. 92-20-85), Puget Crimson (WSU 2833), Puget Reliance, Rainier, Shuksan, Stolo (B.C. 96-33-4),

Sweet Bliss (Orus 2180-1), Totem, Valley Red (ORUS 1790-1); Day-neutral: Albion, Diamante, Monterey, San Andreas, Seascape, Selva.

(More good strawberry varieties for British Columbia, according to the Fraser Valley Strawberry Growers Association). June-bearing: Charm, ORUS 2427-4, Puget Crimson, Sweet Bliss, Sweet Sunrise, Valley Red.

3. Manitoba

(As recommended by Manitoba Agriculture, Food, and Rural Initiatives Crops Knowledge Centre).

June-bearing: Kent, Glooscap; Day-neutral: Seascape; Everbearing: Fort Laramie, Ogallala.

4. New Brunswick

(As recommended by the New Brunswick Department of Agriculture, Aquaculture, and Fisheries).

June-bearing: Annapolis (early), Blomidon (mid-to-late), Bounty (late), Cavendish (mid-season), Glooscap (mid-season), Kent (mid-season), Veestar (early).

5. New Foundland and Labrador

(As recommended by StrawberryPlants.org).

June-bearing: Cavendish, Glooscap, Kent; Day-neutral: Seascape, Tristar; Everbearing: Fort Laramie, Ogallala.

6. Northwest Territories

(As recommended by the Northwest Territories Territorial Farmers Association).

Any Alpine strawberry variety (Fragaria vesca species).

7. Nova Scotia

(As recommended by Atlantic Food and Horticulture Research Centre in Kentville; Horticulture Nova Scotia).

June-bearing: Annapolis, Kent, Mira, Sable.

8. Nunavut

(As recommended by *strawberryplants.org*).

June-bearing: Cavendish, Kent; Day-neutral: Seascape, Tristar.

9. Ontario

(As recommended by the Ontario Ministry of Agriculture and Food).

Early to early mid-season June-bearing: Annapolis, Brunswick, Darselect, Evangeline, Glooscap, Honeoye, Itasca, Mohawk, Sable, V151, Veestar, Wendy; mid-season to late mid-season: Allstar, Cabot, Cavendish, Governor Simcoe, Jewel, Kent, L'Amour, Mira, Sapphire; late-season: L'Authentique Orleans, Serenity, St. Pierre, R14, Valley Sunset; day-neutral: Albion, Evie 2, Seascape.

10. Prince Edward Island

(As recommended by the Prince Edward Island Department of Agriculture and Forestry).

June-bearing: A.C. Valley Sunset, Cabot, Glooscap, Jewel, Mira, Orleans, St. Laurent.

11. Quebec

(As recommended by the Agriculture and Agri-Food Canada (AAFC) sub-station in L'Acadie).

June-bearing: Annapolis, Chambly, Harmonie, Honeoye, Kent, La Clé des Champs, Mira, Saint Laurent, Saint-Pierre, Yamaska; day-neutral: Albion. Suitable strawberry varieties for Quebec according to the Lareault Nursery: June-bearing: Annapolis, Bounty, Cabot, Cavendish, Chambly, Clé des Champs, Évangéline, Flavorfest, Glooscap, Harmonie, Harriot, Honeoye, Jewel, Kent, Lila, Mira, Sable, St-Jean d'Orléans, St-Pierre, Serenity, Sparkle, Summer Dawn, Summer Rose, Summer Ruby, Valley Sunset, Veestar, Wendy; Day-neutral: Albion, Charlotte, Mara des Bois, Monterey, Seascape.

12. Saskatchewan

(As recommended by the University of Saskatchewan Fruit Program).

June-bearing: Annapolis, Bounty, Cavendish, Kent; Day-neutral: Fern, Seascape, Tristar; Everbearing: Fort Laramie, Ogallala.

13. Yukon

(As recommended by Yukon Agriculture Research & Demonstration).

June-bearing: Cavendish, Kent.
(Courtesy of Strawberry Plants.Org).

Chapter 3
Best Time of Year to Plant

Best Time Overall

The best time of year to plant your strawberries is going to be dictated on your climate, whether it's super-hot, cold, or also by the year itself if it's an early or late year. The following are good explanations of the advantages of planting at certain times. The best time to plant is late winter or early spring, just before new growth starts. Plants are still mostly dormant, and they'll be shocked the least, plus they'll have a whole growing season to save up energy for the Winter months ahead.

Late Winter

This time of year is good, as your plants will still be in dormancy (they will not be growing actively). Also, some benefits include:

- Earlier establishment of plants
- Earlier fruit production possible
- Increase in fruit production due to better establishment earlier
- Less chance of plant stress and dying because of hot weather-related issues
- Save yourself time planting now before the busy growing season kicks off

Early Spring

Your plants are perhaps just coming out of dormancy. Buds may start to form, though they are still mostly dormant. This time of year, offers the following benefits:

- Protection for plants, especially in a frigid climate, ensures plants don't get frozen before growing from a late frost.
- Potential sale depending on where you buy them as most nurseries and stores have spring sales just after winter, trying to sell old stock to replenish the new.
- Ease of establishment with a bit of care such as watering is also prime because the soil is still cold, and the days are still short, allowing plants to focus on root growth and top growth without the heat scolding them.

31

- Decent fruit production still as plants will have time to store some good nutrients if fed and appropriately planted (more on this to follow).
- It lessens plant stress due to allowing some time to adapt to the area before late spring and summer set in.
- Weather may be more agreeable to plant now, depending on where you live for planting.

Early Fall

Your plants will have gone through the growing season already and will wind down from all the active season, getting ready to go into dormancy. This time can be an excellent time to plant as long as you take extra precautions, especially if living in a frigid climate, not to allow plants tops and roots to freeze as they'll be newly planted and not established. Some benefits and constraints include:

- Fall sale at local nursery or store due to selling off stock for winter months and upcoming tax season.
- There is ample time for plants to get situated in planting locations.
- If prepped well with mulch/clean straw to protect plants through winter, these can sometimes be the most robust plants as compared to the others.

- More time spent prepping and ensuring rodents don't eat plants, also keeping plants safe from super cold weather, mainly extra mulch coverage will help.
- Least amount of plant stress for the following growing season if you get them through the winter.
- Lessen your load of work for the start of the next year.
- Great fruit production and better establishment if planted and mulched properly at planting, while still applying ample fertilizer in the late winter early spring.

Chapter 4
Soil Conditions Required

Strawberries Preference

Strawberries love a rich, silty loam soil, filled with a decent amount of organic matter and a pH of 6.0 to 7.0. However, the ideal is 6.2 pH (soil pH is a scale for how sweet or sour the soil is. Also, it allows certain nutrients to be absorbed in different levels. Its range is 0 being overly acidic to 14 being too alkaline). Personally, the best-growing strawberries I've ever seen were grown in a 6.5 soil pH, in a sunny location with some sawdust spread evenly across the soil for moisture and nutrient retention. Strawberries also love being in a raised area in a well-drained spot, so situate them on a berm (it's where you increase the soil level for the planting area compared to the standard soil grade next to it). If you create a nice small berm about six inches to one foot above the rest of the soil grade, your plants will love it!

How to check your soil's Water Drainage and Soil Type

Testing Drainage of Site:

Items Required:

- Shovel
- Watering hose/or canteen
- Water
- Tape measure or an equivalent measuring device to approximate length

Soil Test Procedure:

- Dig a 1x1 ft hole approximately 1 ft deep.
- Fill the hole with water (do *not* overfill).
- Allow the water to sit for at least 1 hour to saturate the soil.

    1. Refill the hole with water, this time stopping when filled 2-3 inches from the top of the hole.

2. Allow the hole to drain for a minimum of 1 hour, preferably 2-3 hours, while continually measuring the amount of water drained out of the hole per hour in inches.

After 1-3 hours of hourly measurements, you can compare with standardized values to determine your soil conditions. If the water level in the hole drops:

- Less than 1/2 inch per hour: You have poorly drained soil and should amend the soil for better growing conditions.
- 1/2 inch to 1 inch per hour: You have moderately well-drained soil, which is perfect in most cases for growing strawberries.
- Over 1 inch per hour: You have well-drained soil suitable for strawberries, by adding a little extra organic matter to hold moisture into the ground for these plants so they don't wilt or dry too quickly.

Poor Drainage	Moderate Drainage	Good Drainage
Less than ½ inch water per hour (Worst)-Most work	½ to 1-inch water per hour (Best)-least work	Over 1-inch water per hour (In the Middle)- A little more work

Next Checking Your Soil Type
Clay Types

You can dig into your soil and note if your soil seems very dense, sticking together well, and not draining quick, chances are because you've got lots of clay in your soil. This type of soil is prevalent in lowland bottom areas or near slow-moving waterways. Suppose you've got this soil, plant in a slightly raised bed by working topsoil, or composted steer or horse manure into your soil, about a 2-parts compost with 1-part native soil on your site. This process will help feed your plants and get their roots out of the standing water in the wet months to make sure their roots don't rot!

Sandy Loam

If your soil is a moderately soft structure holding together in chunks, then breaking apart easily, chances are you've got a silty loam soil structure. This soil structure is preferred as you'll need to make the least amount of adjustments to hit that favorite spot. Add a 1 to 1-part soil to composted manure in the soil. You have the option to create a raised bed or not, as drainage isn't a problem. I've found a raised bed is still good in almost all cases as it creates a nutrient-rich smaller area where you can easily keep weeds and other plants from growing in with your plants. This allows you to control the plant's environment better, allowing for more manageable trouble-free growth.

Sandy Soil

Next, if your soil is of a sandy structure not holding together, you'll need to amend your soil with lots of good water holding topsoil and composted manures such as steer or horse manure with rotted hay or alfalfa. You can also use a sawdust mix with manure to save money. The goal here is to hold water into your soil for as long as possible that way. This way, it will not leak out, and your plants wilt and die off from a lack of moisture. You'll need to add preferably a 2 to 3 parts compost in this soil and or composted manure mixed to 1 part of your native soil to achieve proper water retention. Don't cheap out on these amendments as your plants will surely not do as well, and you're only hurting yourself with less harvest, or perhaps losing your plants to extra stress.

Make Your Future Easier

Plant your strawberries in a location near your house so you can avoid having to carry water or loads of compost (especially if you need to buff soil to make richer for planting). Place your plants in an area that already has the above better conditions as much as possible to enable yourself to put in less effort changing soil structure and still have a great harvest. If you have a mostly afternoon sunny spot in a slightly raised area or berm with better drainage and richer soil structure, naturally use this to your advantage, thus saving you more time and money.

Chapter 5
Light Requirements

What Light Allows

Sunlight allows for your strawberry plants to go through a process in which they self-process and create their food called chlorophyll; it creates this through exchanging nutrients, oxygen, water, and carbon, to be exact. This process creates the chlorophyll for the plant, and then the chlorophyll is what the plant uses to absorb sunlight effectively to grow and produce lots of energy in the form of fruit. When you take any portion of sunlight out of this equation, you are disregarding a very critical step, and your plants will undoubtedly produce fewer crops. They will grow extremely poor, lacking full sun exposure. This is the part where your plants don't need sunscreen.

Increase Yield

Increasing your yield for harvest is huge! One must assess their situation by looking at the plant's needs and seeing the light, nutrients, water, and mulch available to your plants. Once you get these items in a balance, you'll reap amazing results, one which will repay the grower multiple times over. Take a quick survey of your garden or yard and see where the sun rises, where it stands midday, and where it sets in the evening. Remember, the sun always rises in the east and sets in the West. From March through June, the sun is highest in the Northern hemisphere, and the opposite is true for the Southern hemisphere, having your highest times from September through December. For those living in the U.S. or Canada, the first will apply to you and for anyone about parallel to our latitude.

Sun Requirements for Strawberries

Strawberries love having full sun (defined as 8 hours a day) of light to produce bountiful crops. This plant requires ample light to bring energy to the plant through photosynthesis and proper ripening for fruit production. Generally, the more sunlight your plants are exposed to, the higher yields and sweeter tasting fruit you'll receive.

Hot Areas

In areas where the heat is a big issue, you'll want to supply later afternoon shade to your plants via placing them on one side of the garden that gets full morning and afternoon sun, with partial sunlight in the hottest afternoon parts of the day. Remember, heat isn't necessarily bad; it just means you'll need to stay on top of your watering schedule, though you'll have amazingly great tasting berries for your efforts as well!

Cooler Areas

In these types of areas, say around coastal climates or areas of the like, you'll want to maximize all the sunlight your plant can get access to by situating them in a spot where they'll get as much sunlight throughout the entire day, especially during the warmest part from mid to later afternoon. Colder areas typically prolong the time it takes for ripening fruit, though it doesn't have to be that way entirely. If you do the above tips plus use the following trick, your harvest will still be robust! You can do your plants some good in these areas by placing next to a concrete base foundation, say the sunny side of your house if your foundation is partly exposed. Placing your plants next to your concrete foundation will allow for them to be naturally warmed up quicker and stay at an ideal temperature longer, plus the concrete will reflect some extra light and heat, directing it toward your plants, enhancing their overall growth.

Chapter 6
Water Requirements

Strawberries Need Water

Strawberry plants need ample amounts of water. Suppose your area isn't getting natural rainfall between 1-2 inches per day in the fruiting and growing seasons factoring in temperature as well. In that case, you'll need to supplement your plants with either traditional style irrigation via hose or sprinkler or drip irrigation. I recommend drip irrigation as you'll save a lot of time and money on your water bill). Strawberries will use a lot of water from the time of flowering through until harvest.

- Fun Fact: Drip irrigation also encourages less leaf disease and fungal problems due to keeping the water out of the canopy of the leaves, thus not creating a

- warm moist environment prone to problems such as mold, mildew, or disease onset.
- Strawberry fruit is comprised of roughly 92% water! No wonder they need lots of water to grow and produce!

Know your Soil's Water Retention Abilities

What're Your Soils Like

Dry to Touch Sandy: If this sounds like your soil, you must incorporate rich moisture retaining composted mulch and manure into your soil and raised beds. Do this, as stated in Chapter 4. This will create a much more ideal environment that will hold water particles for a much longer time.

Dry to Touch Dense Clay: If this one sounds like your soil, it will be a super-hard dense material when dry and super mushy, like a soggy sponge material when wet. This soil is not preferable as it will create an environment where your strawberries roots will get root rot. (Root rot happens when there's an excess of stagnate water, and a lack of oxygen that gets to your plants' roots due to the soil not allowing air particles easily into soil, thus killing your plants from water suffocation and oxygen deprivation). To rectify and fix this, create a raised bed and mix in equal parts of additional new topsoil with composted manures. After that's done, apply a good layer of mulch onto the top of the soil to allow for better water retention and mineral content breakdown into the soil.

Somewhere in the middle: If this sounds most like you, that's a good sign. You'll have to do the least amount of work. Your plants will probably already grow in your soil as is, though to do exceptionally well, you'll want to incorporate well-composted manure still and top-dress with ample mulch to provide good nutrients while still holding adequate moisture in the soil.

Things to Remember

Soil is a science and a skill of art. You'll surely need to figure out what works best in your location, though these practices from above will get you close to the best success. The best way to improve your soil is by annually maintaining and adding more compost and manure to your soil and developing good soil structure. In doing this, you'll always be building up your soil structure and increasing microbial activity. (Mycorrhizal is a fungus that's been around since all plant life and improves nutrient exchange by increasing the mass of a plant's root system). If you put more back into your soil (via compost) than you take out, you'll always be getting a better, more nutrient-rich harvest year after year! (This is usually not a common practice, one which is heavily disregarded today. Even still, it's this practice above, which is the truest, most sustainable way to grow food).

Chapter 7

Mulch and Compost Requirements

Start Off Right

The best sure way to succeed is to start growing right from the get-go. Strawberries love a somewhat slightly acidic, almost neutral pH (6.2), as their mulch should help mimic this. The best mulches to use are numerous and can get very pricy if going fancy. I say keep it simple and budget-friendly! The following are some mulches that you can find in most parts of the country and are relatively inexpensive, though they work extremely well! Types of Mulch: (You'll use these on top of the soil after planting in).

Sawdust

This one usually has a more acidic pH due to its nature of the harvest. It is usually being derived from evergreen trees. If you can get this type from deciduous trees besides oak trees, it will be closer to a neutral to slightly acidic pH.. This type of mulch works exceptionally well for top-dressing plants after planting. Spread evenly over the soil after planting to about 2 inches in depth.

Bark Mulch

This type of mulch is also usually acidic due to being derived from coniferous trees, though if derived from deciduous trees besides oak, it usually is perfect. This mulch type adds far more trace minerals and nutrients to the soil than sawdust due to bark storing all the nutrients for a tree's growth. Apply evenly over the soil, keeping to about 2 inches in depth.

Rotted Straw/Hay

This mulch is rich in organic matter and improves soil fertility well. Caution you'll want to use already well-composted material as its new raw form will usually suck a lot more nitrogen out of the soil than even sawdust or bark mulch. You can also counteract this by adding some nitrogen fertilizer or rich-compost over the top of this mulch to counteract the unbalance without getting on or near plants.

Types of Compost: (You'll use these when planting, mixing them into the soil)

Composted Horse Manure

This, by far, is one of the best as it rarely burns plants at all due to having a lot less ureic nitrogen compared to cow manure. This type is usually not in stores but can easily be found or gotten for free at your local stable or farm. Another significant thing about this type is that it adds a lot of rich organic material to your soil as a horse only digests about halve of their food, so it's also a good mix of composted rich hay and or alfalfa, very rich in nutrients. Mix this well into your soil, following what we discussed earlier for amounts based on soil type; more in Chapter 4 on this.

Composted Steer Manure

Another rich option to use and is just cow manure usually, though called steer manure oftentimes. This type of compost will not burn or hurt plants than its raw uncomposted form due to the high nitrogen ammonia being far less concentrated. Cow manure is a good mixture for soil as it also gets the good soil microbe workers going, which your plants need to survive and thrive. Mix into soil well following Chapter 4 on mixture amounts based on soil types.

High-Quality Topsoil

This is more of an addition to your native soil if it's of terrible material. You'll use this with a 1 to 1 ratio plus adding the above compost to your soil. If your soil is utterly inadequate, you'll want to just stick with new high-quality topsoil mixed with a good compost from above, leaving out your native soil in the mix, then top-dressing as stated with a little mulch in Chapter 4.

Chapter 8
Fertilizer Requirements

What Strawberries Like to Flourish

Strawberries like rich, balanced soil. When planting, you'll want to ensure you're providing your plants with a little extra phosphorous, which will aid in quicker establishment leading to healthier plants faster. However, during the growing period, after plants have gotten more established, usually after a growing season, you'll want your nitrogen to be decently higher than the other three nutrients to facilitate good top growth during the growing months. During fruit production, your plants will use a lot of trace minerals and potash in the soil; therefore, you'll want to provide more of this essential element before the heavy onset of blooms and fruit, thus allowing for better nutrient absorption. You have many options that you can use to give nutrients to your plants. I'll discuss with you some benefits and drawbacks of each. Usually, just one or two of these fertilizers will be sufficient for optimal growth.

How to Grow Best for Success

First-Year of Planting

When planting in late winter/early spring, fertilize with a light balanced fertilizer at planting. If planting in the late winter or early spring, airing on the side of less to burn plants during this sensitive time. As per application, follow instructions on bags, though as a general rule of thumb, if using a 10-10-10 fertilizer, you'll apply one pound of fertilizer per every 100 square feet.

At planting in early fall/early winter, avoid fertilizing as plants will not readily absorb many nutrients due to being in dormancy for a prolonged period. During this time, it's best just to ensure your plant with a suitable soil medium and wait to fertilize with a well-balanced fertilizer until late winter or early spring.

Just after your first harvesting has passed during early summer, it's a good idea to fertilize your plants once more before the growing seasons over. Use a higher nitrogen fertilizer; this will help your plants save up lots of lost energy from the current growing season for next year's harvest. In doing this, you'll allow your plants to store up more energy in their roots for a robust and healthy harvest next spring. Before you do any of this, assess your plants. Are they showing apparent signs of nutrient deficiencies? If growth is severely stunted up top, it could show a lack of nitrogen, which helps promote lush top growth. This is the most common lacking nutrient due to strawberries being heavy nitrogen feeders. Also, ensure plants are still receiving ample amounts of water and ample drainage not to be waterlogged, as discussed earlier, as this can also be an issue with stunted growth.

Caution: Don't over-fertilize your plants with excess nitrogen as it will burn their roots and lead to plant mortality! A simple way to avoid this is to follow the manufacturer's instructions for a balanced application when applying. Take this seriously and take the guesswork out of this by using a measuring device to ensure the proper application is achieved. Too much and your plants could get hurt; too little, and you're setting yourself up for a sub-par harvest to what you could have.

Always air on the side of slightly less if you're unsure, though, as it's better to have your plants alive because of a little lack rather than be burned because of too much. Remember everything in moderation; this is especially true for plants!

Second/Third Year of Planting

Late winter/or early spring: Fertilize with an excellent fertilizer significantly higher in nitrogen and slightly higher in potash, such as a 12-5-7, keeping with application rates specified on the label. Fertilizing with high nitrogen is essential for maximum growth for fruit and plant growth. Potash doesn't usually readily deplete out of the soil as quickly as nitrogen, though it is essential to include it still because it feeds your plants in fruiting production ways. Think of it as a big determining factor in how well your plants will fruit. Plants are likely to use many minerals plus nitrogen and lots of water to produce loads of fruit. Science still debates precisely how this unique element works to do just this. However, one thing is known: many gardeners, including my own, success has been closely linked to the right timing; before the onset of blooms and fruit of high nitrogen, with good potash-rich soil.

Early to Mid-Summer

Just after you harvest all the fruit, fertilize with a high nitrogen fertilizer. The high nitrogen content will allow plants to replenish and still grow well into the early Fall. *Caution*: If you're in a concise growing season area, be careful to allow a little more time until the dormant months from when applying fertilizers as you don't want your plant being encouraged to grow in the winter when it should die back up top and resting for spring.

Timeframes for this will be slightly different based on variety, location, and types of strawberries planted, though no matter if June-bearing (most common), everbearing, or day-neutral, it'll be after the first harvest. For Day-neutral (most uncommon type), fertilize during the first week of July, thus aiding in bloom/fruit production after that.

Benefits of Nitrogen (Nitrogen-P-K)

- This is the most common issue that strawberries lack, getting enough for healthy growth.
- Essential for fruit production as it's a key protein enabler in biomass production.

- Allows for lush leafy-green growth in the whole plant, including helping enable bloom production.
- Quicker growth, avoiding stunted growth.

Plant Indicators of Nitrogen Deficiencies

- Plant leaves very light green and small and then turn yellow during the growing season.
- Purpling Leaves can show a nitrogen issue, also a chance of a phosphorous problem.
- Shortening and smaller leaf petioles as compared to normal, turn a reddish color. Don't mistake the color change during the fall months when leaves die back and sometimes turn a crimson color based on weather.
- Plant's midterm and older growth leaves will show signs first as new leaves will use all available nitrogen, cannibalizing it from older growth. If a nitrogen deficiency is suspected, you can take tissue samples from older growth to have sampled at your local cooperative extension or equivalent to test them.

Organic Sources of Products Naturally High in Nitrogen

- Blood Meal (Approx. 12% nitrogen by content, fast-acting to absorb).
- Fish Meal liquid (Approx. 10% nitrogen by content, fast-acting to absorb).
- Feather Meal (Approx. 15% nitrogen by content, also slower to release).
- Alfalfa Meal (Approx. 3-5% nitrogen by content, a more balanced fertilizer, slower to release overtime).
- Bat Guano (Approx.10% nitrogen by content, fast-acting).
- Chicken Manure (Approx. 3-6% nitrogen by content, fast-acting, caution as it can burn if not composted).

Benefits of phosphorous (N- phosphorous-K)

- It supports vigorous root growth.
- It supports in small ways for fruit production.

Plant Indicators of Phosphorous Deficiencies

- Seldom a problem as this nutrient is a low demand nutrient for strawberries, though if noted, Albinism can occur, resulting in unripened, white, tasteless fruit. Though very rare, it can also occur from a large excess burn of nitrogen in the soil.

- CAUTION- Use only in small amounts as excess can cause water contamination via water runoff into lakes, increasing algal blooms resulting in less oxygenated water and a toxic environment for fish.

Organic Sources of Products Naturally High in Phosphorous

- Bone Meal (Approx. 15-27% phosphorous by content, slow to release error on the side of less than too much will throw off plants' growth easily).
- Bat Guano (Approx. 3% phosphorous by content, quicker to release).
- Crab Meal (Approx. 3-6% average phosphorous by content, slower to release, gardeners, love due to containing a natural chemical called chitin, a natural bug repellant).
- Alfalfa Meal (Approx. 0.5-1% phosphorous by content is a more balanced fertilizer with a slower release).
- Rock phosphate (Approx. 20-33% phosphorous by content, not as uniform for quality as unique

rock formations have different nutrient contents, reasonably slow to act as it's long-term).

Benefits of Potash (N-P-Potash)

- Plants grow healthier, aiding in water and nutrient uptake.
- Plants using less water.
- Less disease and insect-prone.
- Produce more bountiful crops.

Plant Indicators of Potash Deficiencies

- Brown and yellow spots and or discoloration around the edges of the leaves.
- Leaf curling includes iron Chlorosis (usually due to the soil's pH being too high above 6.8).
- Plants seem stunted, even with nitrogen-rich soil additives. This is evidenced by a lack of potash in the soil.

Organic Sources of Products Naturally High in Potash Sources

- Kelp (Seaweed) (Approx. 4-13% potash by content, quicker to release) contains approximately 60 trace minerals as well to boost, though be careful to follow application rates as a little of this goes a long way.
- Hardwood Wood Ash (Approx. 3-7% potash by content is quicker to release small amounts at first). Be careful as this will also increase soil P.H.
- Greensand (Approx. 3% potash by content. It has a slower release since it is not as water-soluble). Be careful as this will also increase soil P.H. (natural element derived from a mining element process containing up to 30 minerals to boost).

How to Read a Fertilizer

All fertilizers have three numbers on the front of them. These numbers are as follows N-P-K. These stand for nitrogen- phosphorous-Potash. Nitrogen promotes lush leafy growth for the plant. In terms of strawberries, they are heavy nitrogen users, and it helps with plant runners and leafy growth. The plant's root system uses phosphorous, and it helps promote large healthy roots for your plants. The plant uses potash for overall plant vitality and absorption of plant minerals and nutrients and aids in stem and fruit development.

Helpful Tip: The most common lacking nutrient in most garden and agriculture soils is nitrogen. Nitrogen is one of the most used nutrients for plant growth; it quickly diffuses into water by most means; therefore, it runs off soil smoothly. A simple soil test will show soil deficiencies and excesses. Therefore, I highly recommend them for new gardeners to get an idea of the general area you're working with and what you'll need to focus on improving in your landscape. A simple way to get one done is ordering a kit online or contacting your local county cooperative extension service and paying a small fee to have them analyze a soil sample of your ground to get an idea of how your soil is composed. These can sometimes be the best means of cost savings as you'll find out if you need to be spending extra on fertilizers and or composts to achieve good plant growth.

I place a strong emphasis on soil culture as nutrients that are out of balance in your soil will surely be noticed by your plants! Be careful when you select a fertilizer to select based on your plant's needs in that type of soil, deciding if you would like to stick to a more natural or chemically derived approach to feeding your plants. The following will list some benefits and drawbacks to both.

Chemically Derived

They make this type of fertilizer from chemical substances, and is not a natural means of feeding your plants. Plants will still register the food as the same as the natural counterpart, though. Chemical types of fertilizers will feed your plant; however, it won't feed your soil, though the plant won't recognize the difference between chemical or naturally derived nutrients. A good article by *Envirolngenuity* reiterates this well, "Plants cannot distinguish between an organic or synthetic fertilizer–the nutrients are processed in the same way. However, the similarity stops there. (*Envirolngenuity*) That being said, your plant will not notice, though your soil will notice. The beneficial organisms living in it, helping aid your plants in growth, will severely take note and could suffer. Tip to keep in mind with this approach is that when you feed your plant, you're not just feeding your plant, you are feeding your soil, and your soils really feeding you. Ask yourself this question in a lawn setting. This isn't as crucial in using unnatural means as you're not eating the grass, hopefully. Though you're going to be consuming this product's fruit, it begs to question using that which has no value to your soil health. Take a skin product. For example, the chemical may work better in some accounts for removing blemishes, though ask yourself if you really want to chance consuming harsh chemicals into your body. The connection can be made that you're not actually consuming these harmful nutrients though I've always realized that if the plant absorbs them and uses them in its growth, they are still a part of that fruit.

Instructions on most of these types recommend wearing masks and glove, which sounds like it may be hard on your soil too? Realize that the same is true for natural means like blood meal, or bone meal, though those recommend it mainly for your skin and avoiding animal by-product issues. Chemical fertilizer is available to the plant the fastest, which is optimal. However, on the flip side, it could also burn the plants a lot easier. It by-passes all the natural processes of decomposition via the microbial activity, which would help the plant consume the nutrients, thus lowering if not killing these beneficial workers in your soil. Always ponder the idea that states, "We are what we eat," and our plants are what they eat in part; therefore, take caution on using this fertilizer if consuming for food. This type has benefits such as quick absorption, ease of application, and good consistent results. It's up to you to decide if it's right for you.

Naturally Derived

Helping your plants and soil out in one simple go by using naturally derived fertilizer is best for good soil to plant equilibrium. When I say natural, I'm not saying you must use organic or anything like that if you don't wish, just use something that is plant or animal-derived rather than that in which is derived from chemical bases. These types of fertilizers are what nature and God intended for people to use. These types come in the form of manures, kelp, blood meal, bone meal, cottonseed meal, a mix of these, or anything derived from plant or animal products. These substances are excess when either processing animals or raising animals via manure or plant-derived like cottonseed meal or kelp. These types of nutrients will feed your soil and your plant sustainably. With natural types, you'll probably notice your plant requires fewer fertilizer applications, less water, and more heat and drought resistance.

The reason behind all the above is quite simple: when you feed your soil, these natural soil building amendments your feeding the healthy microbial activity in your soil and helping promote them to grow and produce, which your plant relies on for a healthy growing environment. Also, naturally derived nutrients usually take more time to become available to your plants fully. This could be good or bad if you need a lot of nutrients quickly, though their useful life once applied is much longer. On the flip side, the chemical counterpart only feeds your plant and often kills these organisms or hinders their essential role due to death by chemical. It's important to note that plants won't create a healthy relationship with mycorrhizal fungi when chemical means are used. These fungi also enable your plants to grow bigger, healthier roots, which allows them to absorb nutrients tenfold. Arguments for both sides can exist; realizing this, it's up to you to decide. Lastly, it's important to note that chemically derived fertilizer came about during the start of large-scale farming as it was easier to produce on a large scale for food and gave a consistent feed to plants; still worth noting though it's lack of long-term soil improvement is a tradeoff for aiding in only plant growth.

Know What's In It! Why You Should Care!

Chemical Types

Derived through "man-made" processes, which usually are derived from the excess petroleum industry, examples of the nutrients are Ammonium Nitrate, Ammonium phosphate, Superphosphate, and Potassium Sulfate. These types are then fortified with some trace minerals for the plant. Remember that plants need more than just pure nutrients to survive, including needing organic matter and living organisms that this type doesn't support. An article of studies by *EnviroIngenuity* states this well "The application of a synthetic fertilizer actually kills a significant percentage of beneficial microorganisms. These tiny creatures are responsible for breaking down organic matter into a stable amendment for improving soil quality and fertility. Some convert nitrogen from the air into a plant useable form." (*EnviroIngenuity*) This is important because these important workers break down all the mulch and compost you apply to your soil, and without them, nothing would get broken down to useable nutrients. Why hinder that? Also, I don't know about you and your families, but I don't want my family to eat a by-product from the chemical or petroleum industry. It's great for cars though I'm not sure about for indirectly ingesting.

We're already consuming enough of our limited supplies, such as plastics and manufacturing. Why use it anymore? Also, these derived types often lower your soil's pH. Another thing to look at is the excess that runs off into water systems, which is another reason most water is becoming more acidic, harming other parts of our water systems. (pH is the scale from 0 to 14 judging how sweet 14, or how sour 0 your soil or water is). Look at this article by Hunker The over-use of chemical fertilizers can lead to soil acidification because of a decrease in organic matter in the soil. Nitrogen applied to fields in large amounts over time damages topsoil, resulting in reduced crop yields. (Hunker). Chemicals have their times and uses, such as when you're not ingesting it like a yard of grass; also, it's more cost-efficient for big-scale farming. From all the previous stated, if you want this for long-term sustainable food production, it shouldn't have that big of a place, especially in your home garden! Grower beware!

Natural Types

They are derived through naturally occurring substances coming from animals, plants, and element formations. These types of nutrients don't rapidly feed your plant. Instead, it releases nutrients a little at the start and then overtime feeding your plant, which is more ideal by reducing excess water pollution runoff, also lowering your fertilizer costs. Microorganisms usually are found in healthy compost-rich soils, in which they convert organic nitrogen into inorganic nitrogen, a process called mineralization. The plants then can absorb this form of mineralization of inorganic nitrogen. Here, you don't cut out the middle-man! :) This middle man works and does all the heavy lifting to ensure healthy nutrients and minerals are available to your plants.

Follow the Old Saying of Getting What You Pay For

When purchasing something that you'll use on a plant and then consume, please be careful, as the goal here is to increase the nutrients in your food and get a better product than the gigantic producers. The Department of Agriculture did a study in an article by Hunker stating, "According to data produced by the U.S. Department of Agriculture Nutrient Data Laboratory, foods grown in soils that were chemically fertilized were found to have less magnesium, potassium, and calcium content." (Hunker) Simply put, the choice is yours; choose to increase your soil's powerful workers in the form of beneficial microbial life with naturally derived nutrients or not. This is a rather deep decision though increasing your fruits' nutritious value would seem most definitely worth it!

Chapter 9
Different Ways You Can Grow

Benefits of Container Growing

- Best if you have minimal space or only have a balcony for growing.
- Ease of movement allows you to live a more mobile lifestyle if you frequently move, allowing you to move your plants easily.
- Avoid rodent problems easily, primarily moles and gophers from eating your plants' roots and killing plants because of being potted above ground.
- Easy Winterization (if in an icy climate that commonly gets below 20-25 degrees Fahrenheit often) by just moving into a warmer space such as the garage or basement area.

- Easier to control soil conditions for abundant growth.
- Easier on your Back as plants can be placed at hands level to avoid being bent over.

Drawbacks to Container Growing

- Less production generally as compared to the other two options. This goes for both fruit and plant production, as plants usually won't have extra space to root new plant runners; plants are generally more water and heat stressed.
- The watering schedule usually is more frequently required because of faster evaporation from the sunlight heating containers quickly.
- Fertilizing is usually more frequent as nutrients are rinsed out of soil medium a lot easier when watering. (This also depends on the soil medium you use, if you use a rich, porous water-retaining medium, this drawback is usually mostly negligible).
- Take extra care winterizing if you have many containers to move, as this becomes timely quick, especially if pots are large or fragile to move, such as terracotta.

- Pots heat up a lot, thus potentially heat stressing plants in summer months much more frequently. If planted in a black plastic container, this happens more often because of the sun's rays heating the plastic and the roots forming a very hot toxic temperature. Overheating can be mitigated by choosing a spot where plants get some afternoon shade or using containers with light reflective colors such as white or brighter colors. This usually won't be too much of an issue if you don't live where temperatures get over 90 degrees Fahrenheit, and you follow frequent watering schedules.

Benefits of Growing in a Raised Bed

- Higher production in plant and fruit production
- More control over soil conditions and drainage issues than direct soil
- Fewer watering requirements - if planted in quality soil and composts, mainly because of increased water holding capacity and less evaporation coming directly from the soil that's well mulched and mixed well
- Easy to keep weeds out - if using quality weedless composts and topsoil's, it's still more manageable

with a nice loose, rich medium making weed pull easy.

- Containment of plants—Much easier keeping plants contained in a designated area, not allowing them to take over an area via runners and duplicating.
- Easier on Back—Allows for less bending over, having an elevated area to care for plants.
- Easier to Stop Rodent problem—Using wire mesh can stop these pests while keeping all the benefits of raised beds. Moles and gophers love raised beds if they are an issue in your area.

  Tip to stop them effectively - Avoid this by using tiny square animal wiring over the bed's bottom, not allowing for entrance into the raised portion of beds. Caution! The wire needs to be less than an inch or two openings not to allow small babies to fit.

Drawbacks of Growing in Raised Beds

- Extra initial costs - raised bed itself, extra soil required, the extra time
- Greater maintenance—Applies to the raised bed itself if constructed from wood due to replacement

costs if rotting overtime, (solution to this could be bricks).

Benefits of Growing Directly in Soil

- Best Maximum Growth potential while growing new plants and berry production.
- Less over-heating possibilities compared to containers, lowering watering requirements.
- Lesser costs upfront associated when you are using less soil and supplies.
- Ease of Winterizing if needed by applying straw or the equivalent on top of plants if you're in a freezing climate is commonly below 20-25 Fahrenheit.
- Least investment in time if soil tests out well for growing strawberries.

Drawbacks of Growing Directly in Soil

- Rodent problems are possible.
- If the soil is inferior, more soil mixing requires more labor to make it suitable for premium growth.
- Low to ground usually making weeding and picking more difficult.

- Possible unfavorable soil conditions are real if the soil is slow to drain or flooding, which suffocates strawberry roots.
- Usually, more weed problems than others, so use an excellent weed barrier properly to solve this.

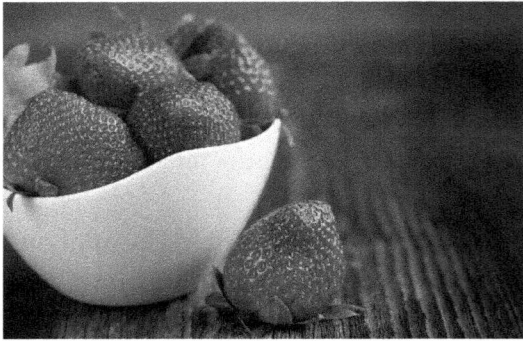

Chapter 10
How to Plant for Success

You'll want to gather up all your tools, plants, and items needed before starting to plant and place near the planting zone to make your life easier, thus allowing for a quick, smooth planting experience.

Tools and Supplies Needed

- Shovel or trowel
- Gloves (somewhat optional)
- Markers (Optional, though recommended if planting more than one variety)
- Bucket and or water hose to water
- Mulch
- Compost

77

- Fertilizer
- Weed barrier (optional, though highly recommended to avoid weed problems).
- Cracked nutshells or ground eggs shells - optional though recommended to keep slugs away. Eggshells contain slow-release calcium, great for plants too. Just don't overdo it!
- Stakes to hold weed barrier down, unless using mulch over top
- A little sweat and hard-work

Steps to Planting

(Steps will be similar if planting in a raised bed, mounded area, or just in the native soil).

1) Put your gloves on, if using.
2) Prepare your site; this is something that was thoroughly discussed in Chapters 4 and 7. Once you've beefed up your existing soil for an ample planting zone for strawberries, you'll need to decide on your plant spacing.
3) The spacing of plants should be done in the following increments: June bearing types, (aka high runner spreading plant varieties plant 18-24 inches from plant to plant, leaving 4 feet space between your rows. If planting plants with less

spreading capabilities such as everbearing/day-neutral plants planting at 6-8 inches apart should suffice, keeping rows 1 foot apart. If container growing, consider planting a couple of different varieties within each pot if size allows for better cross-pollination, also increasing harvest times because of different varieties ripening at slightly different times.

4) Mark areas to be planted keeping with spacing allowances.

5) Dig holes. This step should be easy as you've just incorporated excellent organic material into the soil; if using, the soil will move freely. Dig final size of holes only slightly lower than the depth of the root length on plants, and about half that for the hole's width. Unless if planting container-grown strawberries, then dig 2x the width of the pot, and slightly deeper.

6) If late winter, or early spring, fertilize plants after one month from planting to allow plants first to acclimate and start growing without excess fertilizer, chancing the possibilities of burning them with too much nitrogen. If you are using a naturally derived mild fertilizer amendment low in nitrogen, higher in phosphorous for root growth establishment, you can apply at planting.

Otherwise, apply the designated fertilizer of choice after one month has elapsed.

7) Backfill your amended soil, holding the plant if bareroot while using your shoe or small shovel to slide soil 360 degrees around the plant's roots. Ensure you keep the roots slightly above soil level, primarily if container-grown, to not suffocate roots by planting too deep! Lightly tap soil once filled to compress any air pockets that may have formed. Form a small heap of soil 360 degrees around the plant's root zone, keeping a few inches from the plant's stem to allow water to be held in berm when watering; this will also keep fertilizer from running off as much.

8) Water plant in thoroughly

9) (Optional) Apply weed barrier if using overlapping seams (where pieces meet) at least 6 inches not to allow weeds to grow through. Then cut out a small hole where plants are, place gently over plants, and around them.

10) Apply mulch to the area. Doing this will improve water retention to the soil; this will also secure your weed barrier if using stakes to secure it. If not using a weed barrier, this will still breakdown in the soil and add more great organic soil matter to the soil, thus improving structure, healthy microbes in the soil, and increasing your harvests!

11) (optional) Apply even layer of cracked nut shells or ground sharp eggshells around plants. You'll want to apply to the entire area where plants will grow to mitigate slugs from devouring your fruits and plants. This is highly recommended if you have a slug problem in the least bit. The last thing you want is to do all this work to have your crops eaten by slugs!

12) (optional) Mark plants with garden stake markers or use existing labels if provided from nursery when buying. This will come in handy to allow you to see which varieties do better in your small micro-climate.

13) Done! Now you'll just need to stay on top of your plant care requirements such as watering, fertilizing, mulching, and weeding if needed. Remember, slugs, birds, small rodents, and even neighbor kids will all be a problem in keeping your harvest safe once berries are ready to pick. You can mitigate all these problems by doing the above and placing close to your house for added security.

14) (Optional) Just before berries are ripe, hang bird netting over tops of plants to not allow bird predation if desired.

Chapter 11
How to Protect Plants from Early Frost

Easy Ways to Protect in Spring

During early spring, plants new lush green growth is at its most vulnerable. A sudden frost, once the growing season has already started, can easily set your new plants back quickly if not protected. After planting, before fresh growth starts, place a cutout milk-jug or the equivalent over new plants. This is a super simple process, take a plastic jug, cut the bottom out, then cut a hole in the top for moisture and water to escape. Once placed around new plants, place a small stake inside to secure jug if in a very windy location. Once that's completed using mulch, place about 4 inches up against the jug around its entire perimeter. This mulch and container jug will act as an insulating barrier from the frost, almost entirely eliminating early frost issues. Take note that the mulch will usually be enough to secure the container in place if not in high wind areas.

Doing these steps will ensure your new plants are protected from early frosts if problematic in your area.

Easy Ways to Protect for Fall and Winter

If you're in an icy climate, typically getting well below 25 degrees Fahrenheit, you'll probably want to ensure you insulate plants from harsh chilling. Once the growing season is completed and early Fall arrives, top-dress planting area or container-grown plant coverings with a four to eight-inch layer of straw or sawdust mulch. If you're in a colder climate, it's alright to err on the side of a little more, as this will act as a barrier insulating plants from the chilling frosts.

If grown in containers, place burlap, or bubble-wrap around containers securing with twine or the equivalent, thus ensuring roots in pots don't freeze as bad. Once that's complete, you'll do the same as above, adding 4 to 8-inches of straw or sawdust mulch over the top of plants providing a good insulator for the winter.

Once Frosts are Gone

After the last frost has passed, and before fresh growth emerges, remove all mulch coverings and insulating materials. Plants can now grow safely.

How to Check the Date for Last Frosts in Your Area

An excellent reference to use in the United States is The U.S. Department of Agriculture's interactive Plant Hardiness Zone Map (*planthardiness.ars.usda.gov/ PHZMWeb*). This is a great resource for locating the average annual temperatures in your region. Records will go back since first recordings in your area to give you a wide range of years, taking most of the guesswork out of this. Things to remember here are that the last frost data collected are based on historically averaged temperatures; therefore, there's still a slight chance that a plant-damaging frost can occur after the listed date in your region (although the chance of a killing frost is unlikely). As with most things in gardening, nothing is guaranteed. By following the tips and tricks above, you'll avoid most if not all of this worry.

In Canada, a great reference to use is called The Plant Hardiness Zone Map of Canada. The same can be said about this resource. Nothing is ever 100% guaranteed, as it goes off historical data, and no two years are ever the same. This is still a great resource and can be found by searching (The Plant Hardiness Zone Map of Canada) on the internet or by using this link: *planthardiness.gc.ca*

In all other areas around the world, check with your Department of Agriculture or cooperative extension municipalities if you can't find this information on the internet for your cold hardiness zoning.

Chapter 12
Choosing Between Fruit
or Plant Production

Strawberry Runners

Strawberries are notorious for producing new plants via runners called stolons. These runners come out of the main plant as the plant grows and establishes. One plant can produce 30 to 60 new plants over time! These new plants can be allowed to root into the surrounding soil, and once rooted and growing, you can dig these fresh shoots up and pot them to sell or give them away. Other options are digging these new plants up and making more strawberry beds free from buying more new plants. The downsides to these prolific shoots are that they take a lot of nitrogen for the plant to produce; therefore, this takes away nutrients from fruit production. There can be a happy medium by allowing some of these side shoots to grow and keep producing fruit once they start to grow, allowing more clones of the same plant to grow—more on this to follow.

Fruit Production

Fruit production on strawberries can be immense! Strawberry plants can produce anywhere from half a pound to one pound of rich red fruit annually. Production is highly dictated by strawberry varieties, soil, light conditions, fertilizer, and a good watering schedule. Another factor that can subtract from production is the growth of new plant runners. Things you can do to maximize fruiting is to pinch off these new side shoots as they start to grow from the plants. By pinching or cutting off these shoots close to the plant's base, you'll effectively shift your plants from putting all their energy into growing new plants and instead put that into producing bountiful fruit. Monitor your plants every watering schedule to stay on top of stopping these runners from growing prolifically and taking away from your harvest as fresh shoots will and could grow multiple times weekly!

Happy Medium

I've found that you can usually allow three to five plants to grow off each strawberry plant annually, growing more plants from a single plant and not taking away very much of your fruit production if you take good care of your plants. I love strawberries because you can continue to make new plants for eternity if you continually plant harvest, allowing your old plants to make your new plants for the next few years down the road. This is usually the easiest way to grow new strawberries versus trying to grow from seed. Strawberries grow in containers well if you supply a high-quality soil medium, thus allowing you to keep more plants even if you have no space to plant more at the moment.

Chapter 13
Harvesting and Storing

1st year

During your first year of planting, it's highly recommended you pinch off all or at least half of the blooms on your plants during the flowering season to allow most of your plants' energy to go toward root development and establishment. If you do this, you'll delay some immediate satisfaction, though you'll be rewarded with lovely heavy crops for years after!

General

You'll want to keep a close eye on berry ripening. Depending on your area and variety, selected berries typically are ready to pick within 60 days of planting if planting in spring or very early Summer.

Picking

When harvesting fruit, the best time of day to pick is early morning when the fruit is still cold; this time is best for the harvester, and it's the coldest part of the day. You harvest by pinching fruit off the stems using thumb and forefinger to pinch stem just above the fruit's top, leaving it attached to the fruit.

Note: Care should be taken to not pull excessively on plants as not to disturb plant roots. Ready to pick ripe fruit should pull off the stem with ease. Also, care should be taken to not throw or mush fruit with hands as strawberries bruise easily!

Cleaning/Storing

Once picked, you'll want to wash the fruit lightly off under running water to ensure no dirt or debris are on the produce.

If storing on counter-top: This method is advisable if berries were not fully ripe, as they will ripen more by being in the sun. Care should be taken to consume unfrozen fruit as soon as possible as the shelf life of unfrozen fruit is usually days.

If storing refrigerated: Store fruit in a well-aired container to allow good air circulation not to allow mold to form. This method will keep fruit in excellent condition for up to one week, though it's advisable to eat within three days.

If storing in the freezer: Use freezer bags. Before storing in bags, take off any remaining stems, then place on a baking cookie sheet with parchment paper or foil, spreading fruit evenly across the pan. Place in freezer until frozen. Once froze, take out of the freezer and place into freezer bags. Seal bags well, and store for up to one year. Note: Placing in freezer bags once berries are already frozen eliminates berries clumping or sticking together.

If canning for Jam or Jelly: You'll want to clean berries entirely off and take off stems. Next, you'll want to follow your recipe on canning instructions ASAP. The longer you wait to can from harvest, the greater chance your berries could spoil. It's perfectly fine to allow a day or two to ripen extra on a counter-top with partial sun if desired, though no more than that.

Summary

Before you go, I just wanted to say thank you for purchasing my book. You could have picked from dozens of other books on gardening or the same topic, but you took a chance and chose this one. So, I just want to say a HUGE thanks to you for getting this book and reading all the way to the end.

Now I wanted to ask for a small favor. **Could you please consider posting a review on the platform? Reviews are one of the easiest ways to support the work of independent authors.** This feedback provided will help me continue my mission to write these types of books and continue to help you get the results you want. So, if you enjoyed this book, please let me know! (-:

In conclusion, feel free to always refer back to this book again, as the more times you read it, the more knowledge you'll gain. I wish you and your family the best adventure and the utmost success in growing bountiful strawberries!

About the Author

I grew up on a small farm outside McMinnville, Oregon, located in the Willamette Valley. I learned a lot about myself and how to grow plants successfully. I learned how to grow many different fruit trees, shade trees, windbreak trees, also traditional and non-traditional herb gardens at a very young age. I had many encounters with Mother Nature in the countless crops we grew. Some years were better than others, but by sticking to time-tested methods of propagating plants, I learned better ways to negate most horticulture risks. Hands-on farming was not the only way I learned. I took instructional classes on horticulture, forestry, and outdoor management. When I couldn't get an answer, I would do countless hours of research online from various sources and then implement them into the landscape to determine what worked and what didn't. Today, I bring to you the very best of my knowledge and insight into a wide array of common plant-related issues. I believe that the best way to grow anything successfully is by sticking with what Mother Nature intended. Deviation from this will typically lead to struggles. By reading this guide, I hope you can avoid pitfalls when planting and growing your strawberries.

Happy growing friends!

-John Klein

Sources

Stawberryplants.org: *https://strawberryplants.org/strawberry-nutrition-facts Envirolngenuity:*

http://www.enviroingenuity.com/articles/synthetic-vs-organic-fertilizers.html

Hunker: *https://www.hunker.com/12401292/harmful-effects-of-chemical-fertilizers*

Index

GROW IT! YIELD IT!

Everything on Planting Blueberries for Beginners Success

by <u>John Anthony Klein</u> (Author), <u>Dr. Melissa Caudle</u> (Editor)

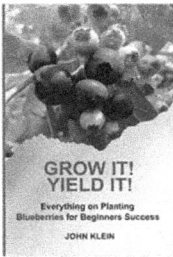

For John Klein, there isn't anything better than picking fresh blueberries, and in his book, he shows how you can grow them in your backyard. This book informs readers with the basic information to not only plant blueberries, but how to yield a plentiful crop so that you can enjoy the freshness of the blueberries and save money at the supermarket not having to purchase an inferior fruit. Easy to follow with beautiful pictures, this book includes:

•Different types of blueberries
•Soil consideration and the best place to plant
•Mulching
•Watering
•Fertilization
•When to plant
•How to keep a healthy plant by pruning
•And a lot more.

<u>Purchase your copy today</u> and learn how to grow your own blueberries.

Successful Growing

by John Klein (Author), Dr. Melissa Caudle (Editor)

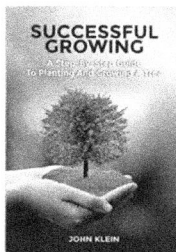

This guide provides high-quality information that people can easily follow to plant and grow a tree successfully. If you are interested in any of the following:

•Learning how to grow a tree.

•Saving money on energy, heating, and cooling costs.

•Making a memory with family and friends

•Enjoyment for years to come

•Saving the planet

•Increasing knowledge to pass on

•Gaining a green thumb

•Enjoying the thrill of watching your tree grow strong and healthy.

The following content can then help you achieve these with thorough instructions and helpful tips on how to plant and grow a healthy tree of your own.

John Klein

GROW IT! YIELD IT!
Everything on Planting Rhubarb for Beginners Success

by <u>John Anthony Klein</u> (**Author**), <u>Dr. Melissa Caudle</u> (Editor

For John Klein, there isn't anything better than growing fresh rhubarb, and in his book, he shows how you can grow it in your backyard. This book informs readers with the basic information to not only plant rhubarb, but how to yield a plentiful crop so that you can enjoy the freshness of the rhubarb and save money at the supermarket not having to purchase an inferior fruit. Easy to follow with beautiful pictures, this book includes:

- Different types of rhubarb
- Soil consideration and the best place to plant
- Mulching
- Watering
- Fertilization
- When to plant
- How to produce more plants
- And a lot more.

Purchase your copy today and learn how to grow your own rhubarb.

GROW IT! YIELD IT!
Everything on Planting Strawberries for Beginner's Success